Standing Room Only

After graduating in applied chemistry Douglas Ashmead spent some time in biochemical research in the coal and textile industries, before moving into industrial management in the metallurgical and nuclear fields. This was followed by many years as senior partner with a leading international consultancy company.

Some twenty-five years ago the author decided to practise as an independent consultant specializing in 'company doctoring'. He also took on the role of part-time executive director of some of the largest engineering and manufacturing companies. More recently he has been doing work in northern Russia on the corrosion protection of major oil pipelines.

Presently he is chairman of electroplating and galvanizing companies which he set up with others in the north of Scotland.

Over many years the author was also a Director of Aberdeen University and of the Edinburgh Language Foundation which specializes in teaching English to foreign students.

Douglas Ashmead is a voracious reader and keen yachtsman. He is married with two grown-up children.

Standing Room Only

Our Overcrowded Planet

DOUGLAS ASHMEAD

ROBERT HALE · LONDON

© *Douglas Ashmead 1997*
First published in Great Britain 1997

hardback ISBN 0 7090 6039 4
paperback ISBN 0 7090 6042 4

Robert Hale Limited
Clerkenwell House
Clerkenwell Green
London EC1R 0HT

2   4   6   8   10   9   7   5   3   1

Photoset in North Wales by
Derek Doyle & Associates, Mold, Flintshire.
Printed in Great Britain by
St Edmundsbury Press Limited, Bury St Edmunds
and bound by WBC Book Manufacturers Limited, Bridgend.

Contents

Dedication

To all my family and friends who have endured my pontification on this subject over far too many years.

The author wishes to record his indebtedness to all his sources, some only half-remembered, who have contributed to this digest.

Thanks are due to several individuals, experts in their own field, who reviewed my earlier drafts and a specific mention is befitting to my researcher Neil Spicer of Glasgow University.

The author alone bears responsibility for the views advanced and for any factual errors that the book may contain.

Preface

I believe the growth of the world's population to be the biggest single concern facing the human race today. Not only that, but its resolution must be made a prime political objective of the West.

In this short book I try to draw together, in digestible form, all the strands of the controversy surrounding world population growth. I invite you to picture the hub of a wheel. The hub represents anxiety about over-population, and it is buttressed in position by a series of spokes, each representing one facet of the problem. This publication will demonstrate that even if you damage or break one or two of the spokes, the hub will nevertheless stay in place.

All too often, discussion about whether or not we should worry about over-population is focused only on food. As long ago as 1798, the clergyman Thomas Malthus argued that the population was increasing faster than the means of subsistence, and that its increase should be restrained. His thesis was certainly wrong in timing – by about 200 years – but absolutely right in substance.

Many excellent, though lengthy, books have been written about the population explosion; too many, you might think. Few are read. Out of 16,000 students at Glasgow University only some 500 in any year think it worthwhile to read even one of the 2,000 books on offer.

To try to overcome this disinclination even to *consider*

reading a book concerned with over-population, this one has been structured to be read in an hour or so. So it has not been designed with an eye to the *cognoscenti* or the professional in his own field. To achieve brevity, this slim volume deals with each facet of the debate and suggests approximate conclusions, rather than covering the full and fine detail of any particular argument. I can, however, be confident that the general thrust of each case is true – but, if perchance you disagree, then you can none the less consider whether or not the hub of the wheel has stayed in place.

This book is written for the reader in the Western World because, even though we have relatively low population growth, we constitute the biggest threat to the survival of mankind through our massive consumption of natural resources. Accordingly, it is we who can make the biggest remedial impact, and it is we with our sophisticated communications systems who can do most to place the topic centre stage and take significant action to mitigate its effects. Even if the book could be made available to people in the Third World, it would be unlikely to provoke much interest. There, the day-to-day pressing needs of survival inescapably limit understanding and concern.

It is important that the subject of over-population becomes the number one concern of us all. Politicians in a democracy, short-term in their outlook and of necessity xenophobic, will only tackle the problem if an informed population pressurizes them to do so. People must be persuaded to believe that the population bomb is a much greater threat than the nuclear one. The metaphor may be questioned because the world will not cease with one big population bang – but it will instead suffer many debilitating illnesses and pestilences such as famine, strife over resources, floods, harvest failures and so on, which collectively would be terminal to our present western way of life.

I should like to re-emphasize that this book is designed to be accessible to a wide readership. Professionals and scholars in demography, world resources and suchlike will not find here any original research in their disciplines. The presentation is an overview, so is not microscopic in nature. I do not know the answer to the question of over-population, but of one thing I am certain – if we do not keep addressing the right question we will not reach the right answer. I do not apologize for anticipating this homily here in this preface.

Douglas Ashmead
Helensburgh
Dunbartonshire

1 How Many of Us Are There?

The world's population is a staggering 5,750,000,000 – less overwhelming to say 5¾ billion – and rising.

Television inures us so much to large numbers that their impact is lost. We talk of international economic matters in billions; of mega-bombs and mega-death; of light years; and of molecules a billion times smaller than a pea. But such concepts are really beyond human comprehension – indeed, even the numbers-specialist finds he is dealing with abstraction. Large numbers such as five and three quarter billion no longer have any real impact.

Does it help to know that 5¾ billion miles is 10 thousand times the distance from here to the moon and back. Does it help if you know that 5¾ billion people laid end to end would go around the world 300 times? No, not really. To most of us, it is just an overwhelmingly large number.

What are the economics of all these people? In 1994, there were 207 countries in the world, categorized by economists into three groups: the Underdeveloped World, the Developing World and the Developed World. In embracing these categories, we have adopted this same simple generalization, ensuring that we grasp no more than the essential underlying argument without distracting detail; the telescopic horizon is what we seek rather than the microscopic one. The Developed World, sometimes known as the First World or the Western World, includes the United States of America, the European Community, Australasia, Japan and the Middle East oil states. It has a population of around a billion and an average income per head of some $17,000.

The Second World or Developing World, including the former Soviet Union and eastern Europe, has a population of some 0.75 billion and a per capita income of $2,000.

The Third World – the 'Underdeveloped' World – includes China, India, Pakistan, Bangladesh, most of Africa and large parts of Latin America. Its population is 4 billion, and the average income is a mere $500.

The significance of these figures is plain to see. The Western World with only one quarter of the total population consumes eight times that of the Under-developed World and on a per capita basis the ratio is more like thirty to one. We are assuming in this short book that the Developing World of 0.75 billion is at economic take-off or beyond.

The average per capita income of the whole world is $3,600. If income was to be equally shared – a highly improbable scenario, though championed by many – we in the West would have to reduce our average income to a fifth of what it is today. Think what that would mean to you! But the solution to population pressure does not rest on even distribution of income. After all, when it comes to making any significant sacrifice, politicians have found out that approval is restricted to the 'they ought to do something' school of response!

Looking from another unreal, impractical (though illuminating) perspective, the world has a predicted economic growth of 3 per cent per annum. If that were entirely channelled to the Underdeveloped World, it would take around 100 years before the West's present standard of living were achieved. But unhappily, the world's population would have quadrupled by then and there is absolutely no reason to believe that the people of the West would be willing to forgo their ambition for ever-increasing prosperity.

However, the prime purpose in writing this book is not to address the difficulties of economic development, or to illustrate the consequences of its achievement by way of increased pollution, global warming, ozone layer depletion, deforestation and more generally the restrictions and restraints on the general quality of life of mankind, particularly in the West. Rather it is to focus more attention than is given at present on the growth of population and economic and other consequences that thereby flow. The hope that the target doubling of the world's economy over the next 25 years can be translated into higher personal living standards is heavily qualified by the expected increase of 50 per cent in the world population over the same period.

2 How Many More of Us Are There Each Year?

This year, an *extra* 100 million babies will be born into the world: in other words, 275,000 extra mouths today, tomorrow, the day after that, and every day after that. If this level is maintained, the world's population will increase by 50 per cent over the next 25 years and *double* by the middle of the twenty-first century.

Human numbers are spiralling out of control. If proof is needed, consider this: it took all of human history to reach the first billion people on earth; the second billion arrived in 130 years, the third billion in 30 years, the fourth in 15 years, the fifth in 12 years – and the sixth in just 11 years. Since the Gulf War in 1991, which seems only yesterday, the population has increased by half a billion – not half a million but 500 million. And each addition needs food, clothing and shelter.

Only humans reproduce this way; the animal kingdom has more sense. While our numbers have doubled over the last 40 years, the number of animals in the wild has reduced by a factor of 4. It is astonishing to think that at the time of Napoleon, 200 years ago, the number of buffalo and wildebeest roaming the great plains of the world was the same as the then human population – 1 billion. We now vastly outnumber all other significant mammals, and we are probably the only unendangered species – assuming, of course, that we do not destroy ourselves.

How rapid is our growth? Does it help to imagine an army of people marching past you three abreast? You would never see the end of the line because the numbers being added equal the numbers marching past every second, every hour and every day of every year.

Does it help to reflect that in the two great World Wars some 70 million lives were lost? The population increase in only one year far exceeds this loss.

Does it help if we say the increase is equivalent to the population of two Great Britains each year? If we tried to give them all the same education as British children we would require something like 300,000 extra schools each and every year with something like 3 million extra teachers each year and 5,000 extra teacher training colleges each year.

One hundred million extra mouths is equivalent to ten Calcuttas; think ten more Calcuttas each and every year; and this would be the more likely outcome. We, of course, recognize the absurdity of such comparisons but, in a way, they do give some feel for scale.

During the minute or so it takes to read this page, another 200 babies will have joined the planet.

The table below shows how the one hundred million extra mouths are distributed.

	Increase in Numbers Each Year (Million)	Children per Woman
Western World	5	About 2
Developing World	15	3.0
Underdeveloped World	80	4.0

There is a popular misconception in the West that our population is falling. As the above table shows, and even taking into account population flows – immigration and emigration – this is just not true. In 1994, the population of the United Kingdom grew by 100,000, and that of the European Union by 700,000, and that of the United States by 2 million.

It is a simple fact of life that if births exceed deaths, the population will increase. This is happening at present – there are around 150 million births but only 50 million deaths a year. This is largely due to rising health standards resulting in fewer infant deaths and people living longer. So it's not that we have started breeding like rabbits; it's just that we have stopped dying like flies.

Over the last 20 years, world-wide life expectancy at birth has risen from 55 years to 65. During the same period the fertility rate has actually fallen, from 5 children per female to 3, but at a slower rate – so the population continues to increase.

Given that health standards will continue to rise – and that one third of the population is under fifteen years old and will soon be of childbearing age – the inescapable arithmetic demands that we must aim for fewer than two children per female across the world.

The need for action is urgent. While the Underdeveloped World must seek the greater fall in birthrate, the developed world's higher rate of resource consumption must also be tamed. Present trends indicate that the number of children per female will reduce to 2 some 25 years from now; a disastrously low rate of change as by then the population will have increased by 50 per cent.

Three quarters of the increase in numbers occurs in as few as twenty of the less developed countries. Africa has by far the highest fertility rate of more than six per woman; Rwanda has eight. But we should not blame everything on the Second and Third World. The West's population is still increasing – and since we consume, and therefore pollute, thirty times more per capita than people in the Third World, they in turn could perversely claim to be entitled to thirty times the rate of growth in population of the West!

In any population equation must come the matter of religion. The following table shows figures for the various religious denominations treating China and Russia as separate non-secular populations.

Religion	Present Numbers (Billions)	Increase in Numbers Each Year (Millions)
Islam (Muslim)	1	30
Catholic	1	20
Protestant	3/4	10
China	1¼	12
Russia	½	8
Hindu	3/4	16
Buddhist	½	4
Total	5¾	100

Muslims account for the biggest increase in births each year, both in percentage and numbers. The principal Muslim countries include the Middle East states, Bangladesh, Indonesia, Iran and Pakistan. Apart from the Oil States they are generally economically poor. Muslims believe that having children or not is the will of God.

The Catholic religion accounts for some 20 per cent of the

annual increase in population, and embraces wide differences in population growth among its followers. Italy, the cradle of the Church, is down to 1.8 children per woman while Latin America has a rate of 3.5. Broadly, contraception is a matter of personal conscience for Catholics, though a firm stance is taken against abortion.

Russia and the former Soviet states have little dogma in respect of population control, and in the present state of unrest I simply cannot see a policy of population control having any priority.

China, on the other hand, has taken a very firm stance and is endeavouring not only to stop population growth but also to reduce it by restricting the number of children to one per family and raising the legal age of marriage. These policies are working (the latter having had much more effect than the former), insofar as the population growth has halved from 25 million per annum to 12 million, but they have many imperfections. Not least among them is the bias against girls, which has reached such proportions that infanticide is practised. The Chinese government believes that, even with its serious shortcomings, its policy is far less appalling than the consequences of uncontrolled population growth, with all the attendant famines and other horrors.

The 16 million annual increase in the Hindu world is of the same order as that of the Catholic faith. The Hindu religion embraces many beliefs ranging from one god to many gods to none at all. Dogma about life is much more unimportant than the flow of life through many existences. The gift of children is determined more by economic needs to support the family than to any religious conviction. The principal country practising Hinduism is India.

Care must be taken in interpreting these figures not to assume that religion is the only cause of high or low levels of growth; other factors such as the stage of economic development are also significant.

In 25 years' time the world's population will be 8½ billion – an increase of about 50 per cent, or just under 2 per cent a year. The increase in the European Union will be some 20 million. This is not some massive upsurge in the distant future beyond our comprehension or concern. If you are under 50 years old, you and your children will experience the consequences of the world population explosion.

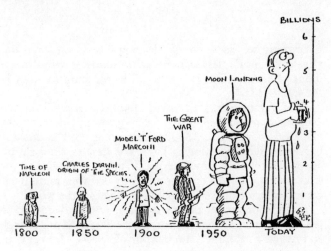

3 What About Fuel Reserves?

Everything we do – eating, sleeping, breathing, even the physically motionless activity of thinking – requires an input of energy. This is true of all living things; energy is the staff of life.

To generate electrical power requires energy. To move vehicles, to make steel or plastics, to grow and harvest food all require energy. All manufacturing and industrial processes require energy. To make deserts flourish by supplying water requires energy. To make fresh water from sea water requires energy. Energy is the underlying common factor of everything that happens in the world.

Apart from nuclear electricity and a very small amount of geological energy from the core of the earth, all our energy sources come from the sun but our direct use of this incredible source is minimal and largely confined to space heating.

To an overwhelming extent, we use the sun's energy trapped in fossil fuels – coal, oil and so on – which are burned for the benefit of mankind. In effect, these constitute the powerhouse created by the sun in the distant past, locked in chemical form. In contrast, today's output from the sun is exploited via biomass, wind and wave sources and a negligibly small amount of direct electrical generation using photoelectric cells.

The sun's energy is trapped by plants, which convert moisture, carbon dioxide in the air and some minerals from the soil into complex chemical compounds. These can then be converted into useful energy by burning. Put simply, carbon dioxide is locked away in plants, timber and so on by a process called photosynthesis, and is released back into the atmosphere when burning takes place.

High-energy fossil fuel sources such as coal, gas and oil are consumed by burning – usually for electrical generation or via the internal combustion engine. Some fuel is used directly for cooking and heating, particularly in less developed countries.

Oil is converted into petrol for use in the internal combustion engine – essentially a burning process. It is also converted into products such as plastics, which eventually degrade. This is fundamentally the same chemical process as burning.

Fossil fuels are generally the end products of trees, plants and very small animals which have been converted under geological pressure over thousands of years. Burn these products and you release carbon dioxide back into the atmosphere. Over the 200 years or so that humans have been plundering these resources, we have been discharging back into the atmosphere the locked-away carbon dioxide of the distant past and causing global warming (see chapter 4).

If we could limit ourselves to using the sun's energy on a current year basis, using only this year's sunshine for our needs, then the carbon dioxide increase and consequent earth warming, discussed in the next chapter, would not occur. The total energy from the sun, if it could be harnessed, would easily meet all mankind's needs many times over – the big problem is how to trap it. Ultimately, this requires trying to collect the sun's energy over very large geographical areas. Photoelectric cells are not really practical on a large scale, so any thoughts about using the sun's energy on a current year basis must depend on 'biomass' – the use of rapidly-growing crops, which can be burned to useful human purpose. This is being urgently explored.

Another of today's energy sources is the motion of waves on the surface of the sea. These are created by atmospheric winds, which originate in the heat of the sun (as opposed to tidal movements, which are caused by the earth's rotation and are independent of the sun).

Developing any of these current year to year strategies, such as biomass, wave energy or wind energy, would take a very long time to have any impact on mankind, and, as you are no doubt aware by now, one of the central themes of this book is that while many ideas are technologically feasible the time-scale of their industrial development is such that the rate of population growth and demand for higher living standards overwhelms the rate at which the likely benefits can be realized.

In making calculations about worldwide reserves and consumption it is necessary to compare the various energy sources in some way that creates a common denominator. This is done by measuring the intrinsic heat content of the fuel – its calorific value – and making allowances for the efficiency of its conversion to usable energy. The common base I am using is the 'ton of oil equivalent' – the amount of oil equivalent to the energy contained within the fuel concerned. For example, 1½ tons of coal is roughly equivalent to one ton of oil. Fuels excluded are wood, peat and animal waste which, although important in many countries, are unreliably documented in terms of supply and consumption.

Total world energy consumption is 8,000 million tons of oil equivalent per year, of which 40 per cent is oil based, 20 per cent gas, 27 per cent coal, 6 per cent nuclear, and 7 per cent hydro-electric and renewables.

Energy consumption in the Developed World is the equivalent of 4.4 tons of oil equivalent per person. In the Developing World it is 1.9 tons, and in the Underdeveloped World it is a comparatively modest 0.4 tons. There are, of course, wide variations: the figure for the USA, for instance, is about 50 per cent higher than that for Europe.

At today's rate of consumption, known world oil reserves will last some 40 years or so. Gas reserves will last another 20 years beyond that, and coal 200 years. But this assumes today's economic activity and population. Energy demand has doubled in the last half century, and if world economic activity increased by 3 per cent per annum, as is forecast, energy demand will more than double in the next 25 years.

However, continuing oil discoveries, for example in the north Atlantic and Far East, may mean that estimates of fuel reserves will be revised upwards. Add to this the variety of new technologies that are becoming available, and things begin to look better. But do not be fooled! Rising living standards combined with population growth will not only overwhelm the reserves in the fossil fuel bank, but also the new technological developments. Even with new discoveries, oil is likely to run out in your lifetime if you are under 40.

Fossil fuel resources are not equitably placed across the planet. Some three quarters of oil reserves are in the Middle East, two thirds of gas reserves are in Russia, the Middle East and the United States, and Russia and China hold as much as two thirds of the world's coal reserves.

The energy consumed per person in the Underdeveloped World would have to be increased 10-fold before it would match the Western World. This is equivalent to a 40-fold increase in total energy demand taking into account total population numbers; and this frightening statistic takes no account of the extra demand that would be brought about by the population increase over the next 25 years.

How will we cope with such a massive increase in demand? To a substantial extent, the expansion of western economies has been dependent on the growth of electrical power generation. Electrical power consumption in western Europe is some 10,000 units per person per year, including domestic, commercial and industrial demand, compared with around 400 units in the Underdeveloped World (a unit being the heat generated by a one-bar electric fire in one hour).

A nonsense but illustrative calculation tells us that to bring the Underdeveloped World up to Western standards would require the overnight building of 4000 very large new power stations. Also, just to keep up with the increase in the world's population of 100 million per year would require the building of about 100 large new power stations each and every year – a next to impossible target if you consider that it takes 10 years to build one power station from concept to launch. Not only that, but there is only a handful of companies today that can design and build such massive industrial complexes.

So what are the alternatives? Renewable sources including wind power, hydroelectric generation, energy from waste and the growing and burning of biomass. Then there is atomic energy. Nuclear energy by fission processes – the present method – is well proven but is accompanied by the perceived hazard of radioactive contamination. This could also yet be true of fusion, the other possible atomic process yet to be successfully developed.

If we could generate power by using fusion processes, we would, for all practical purposes, have an infinite source of energy. But what of the contamination risks? The correct question is rarely posed. It is not the abstract question 'is nuclear energy a good thing?', but rather 'do we want the risk of a nuclear incident as exemplified at Chernobyl, or do we prefer the more invidious risk of global warming from fossil fuels?'

In fairness to the nuclear industry, the number who died directly as a result of the Chernobyl incident is very small, but the concern is about statistical deaths – those caused by cancer and other diseases in subsequent years attributable to radioactive fall-out. Estimates from 1,000 to 10,000 per year have been made. But they should be compared with the environmental impact of other energy sources such as coal and oil.

We need to seek greater efficiency from new power generation systems, find new fossil fuel sources and develop renewable energy sources. Even if all that were to happen, the rate of achievement of these technologies and their industrial development is unlikely to match the growth in population and economic expectations. The outlook is bleak, and failure certain, if we put our efforts solely into meeting the demand of increased population numbers and expectations of higher standards of living. More emphasis must be placed on finding ways to reduce population numbers; growth must be brought to zero and even reversed. But I am perhaps in danger of being the minister in the pulpit preaching against sin. It is easy to preach; deeds are more difficult.

4   Is Global Warming a Threat?

In the main, the requirement to supply heat and food for the human body and to run the machinery of industry, commerce and agriculture comes from fossil fuels, all of which discharge carbon dioxide and other gases into the atmosphere. The effect of these gases on the planet's environment is clearly a matter of concern – particularly the phenomenon that has come to be known as the 'greenhouse effect'.

Most of us have experienced the unpleasant effects of atmospheric pollution, especially in the major towns and cities where nearly half the world's population lives. Smog caused by the motor car and other industrial processes is now a major worry because of the associated health hazards of asthma, cancer and other illnesses. Necessary and urgent steps are being considered to try and reduce this serious nuisance: in some cities, access to vehicles is being restricted to help reduce traffic chaos – and fuel prices are being increased.

But cities are places where people congregate to work – and work requires energy. Coal, oil and gas are needed to run the machinery of industry and commerce and to supply food and heat for the human body.

Clean Air Acts over past decades have focused in the main on reducing the hazardous content of fumes from incomplete burning processes such as vehicle exhaust

emissions. But very often, these very cleaning-up processes have actually created *more* carbon dioxide than before!

What is the greenhouse effect? When light from the sun reaches our planet, it is partly reflected away from the earth, as if it were a mirror. The remainder is absorbed by the earth's surface, which warms in the process. The heated-up planet then radiates the energy it has received back into space by means of invisible infra-red radiation. If the radiation energy – sunlight – coming in equals the radiation energy being given off by the earth, the temperature will stay roughly constant. But if the composition of the atmosphere changes – because of the discharge of carbon dioxide, for example – a blanket effect develops around the earth, preventing some of the heat from leaving. The result is a rise in temperature on the earth's surface, in the oceans and in the atmosphere.

This is called the 'greenhouse effect' after the similar process that occurs in a garden greenhouse on a sunny day when temperatures inside become much higher than outside. The greenhouse glass prevents the incoming heat from escaping.

Until the intrusion of man and his energy-gobbling activities, a natural atmospheric balance was maintained. But no longer. There is considerable evidence that the greenhouse effect is seriously damaging our planet, and the scientific community's concern is not confined to a few people but to scientists in general. Their alarm is of such magnitude that the United Nations held a convention on climate change at which 153 countries contracted to try and mitigate the effects of greenhouse gases and to reduce their production.

Are we right to be alarmed? Is there any evidence that the average temperature of the earth is increasing? The answer to this depends on the time horizon we choose. Certain deductions made from ice and mud core samples at the poles and from study of rocks indicate that over very long time-scales – 10,000 years or more – the earth has experienced Ice Ages, mainly caused by changes in the tilt of the planet's axis. (The best prediction of the next one is 5,000 years from now.)

On the shorter time-scale of hundreds of years the natural variation in atmospheric temperature can be quite considerable. In the past hundred years, for example, the River Thames froze over and there were icebergs in the North Sea. A thousand years ago there was a warm period when vineyards in southern England were not the exception they are today.

There is considerable evidence that real global warming of about 1°C has occurred over the past 100 years: this may sound trivial but the effects are massive. Mountain glaciers have retreated, the global sea level has risen by between four and six inches since 1900, and recent surveys in the polar regions have shown very large increases in the number and variety of plants growing there, indicating a less hostile climate than previously. The inescapable conclusion is that the global temperature has been, and is, rising. This is now received wisdom in the world of science.

The main greenhouse gases are carbon dioxide, methane and chloro-fluoro-carbons (CFCs). Water vapour also has an important role because any atmospheric warming increases surface evaporation, which in turn enhances the warming effect even further; a sort of self-perpetuating escalator mechanism.

Carbon dioxide is exchanged naturally between the atmosphere, the oceans and the living world. About 1,000,000 million tons of carbon dioxide are moved back and forward each year as the process of growth and decay of living organisms on land and in the sea takes place. This balance has been upset by humans, who contribute an additional 25,000 million tons a year.

Carbon dioxide has the greatest effect on global warming simply because it is released in massive amounts – though on a unit weight basis, it is many thousand times less potent than some other gases. The concentration of the gas has increased by 25 per cent since the Industrial Revolution and is presently increasing at about 0.5 per cent per annum.

Carbon dioxide is absorbed by trees and other vegetation. It therefore follows that the more plant growth we have on earth, the more carbon dioxide will be removed from the atmosphere; though in the process of decay the gas is released back into it.

Yet what are we doing to our rain forests and other great areas of greenery? There is a double consequence of cutting down trees in that they are no longer there to absorb CO_2 and when they are felled and burned, the gas they have already absorbed – often over generations – is returned to the atmosphere. Deforestation causes up to 15 per cent of the increase in carbon dioxide.

Methane occurs at a concentration of only about two parts per million, but relatively it has twenty times the effect of carbon dioxide on the atmospheric temperature. Changing agricultural practices in order to feed large populations, together with the increase in waste disposal, has resulted in the concentration of this gas increasing. The massive scale of mining for coal and minerals has added to the problem. Methane eventually degrades to carbon dioxide. Meanwhile, its concentration is increasing by about 1 per cent per annum.

CFCs are man-made compounds. Their non-toxicity and inertness make them suitable for propellants in aerosols, coolants in refrigerators and in the manufacture of plastic foam. Quite apart from their 'greenhouse effect', evidence emerged in the mid-1980s that they also cause depletion of the ozone layer. This seriously concerns all life on earth, particularly humans, because thinning the ozone layer results in more ultraviolet rays reaching the earth's surface. Not least among the consequences is the heightened risk of cancer.

Some international action has been agreed to reduce the use of CFCs – most importantly the signing of the Montreal Protocol in 1987 by more than fifty countries. This agreement pledged to reduce the release of certain types of CFCs by 20 per cent by 1994, and by a further 30 per cent by the end of the century. The European Union has taken the Protocol one important step further, by working towards a complete CFC ban by 2000.

Greenhouse gases produced by human activity are estimated to have the following origins:

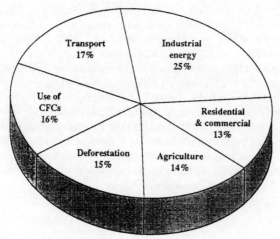

and by geographic area:

Area	% Greenhouse Gases	Population Percent of the World
USA	27	4
Other Developed countries	17	9
The former USSR	20	4
European Union	17	4
China	8	22
Developing countries	11	57

The contrast is large: China with 22 per cent of the world's population produces only 8 per cent of greenhouse gases. America, on the other hand, accounts for only 4 per cent of the population, yet produces 27 per cent of gases.

The United Nations International Panel on Climate Change predicts that if the world takes no special action to limit emissions of greenhouse gases, the global mean temperature could increase five times faster than in the last century. This would result in an increase in global mean temperatures of about 1°C above its present value in about 25 years and 3°C by the end of the next century; a rate 25 times faster than that over the last 2 millennia.

The trouble is that the increase in greenhouse gases is invidious and never dramatic in its short-term effect. When its consequences become manifestly apparent, it will almost certainly be too late to do something about it. It is imperative that we act now.

5   The Effects of Warming

The cardinal effects of warming are a rise in sea level and changes in climate.

The sea level will rise because the oceans will expand and land ice sheets will melt. There are many uncertainties, but best predictions indicate a rise of about half an inch a year, though there will be significant regional variations. Some pessimistic estimates put the rise over the next 100 years as high as 15 feet.

Two thirds of the earth's surface is covered by oceans. Between the seas and the land is the coastal zone, some 15 per cent of land surface, where three quarters of the world's population live and work. Any change in sea level would have particularly serious consequences here, including loss of living space and food growing areas, fresh water problems, difficulties with harbour operations and a fall-off of tourism. Areas at risk today include the Maldives, Bangladesh, and the coastal plains and deltas of China. The Netherlands and the United States also have cause for serious concern, but at least they can afford to take effective action.

It is estimated that 50 million people experience serious flooding each and every year. This number will double over the next 25 years. Think of the already frequent and disastrous floods in Bangladesh and China, and the devastating effect they have on human life. We've hardly

seen anything yet.

Changing weather patterns, including increases in the frequency and severity of storms, could affect as many as a further 400 million people. Think of the flooding over large areas of Europe in 1995 and China in 1996.

The global cost for immediate but basic protection measures is estimated to be in excess of £1,000 billion: roughly the size of the Gross National Product (GNP) of Great Britain. For many countries – especially small island states in the Pacific and Indian Oceans – their share would constitute a substantial part of their GNP. The magnitude of the problem can be seen if you consider that it would require all the output of all the people in the United Kingdom over a whole year to finance basic initial measures to protect the world from floods.

The loss of the planet's coastal land area is already serious and exceeding 1 per cent a year in many places. Nearly 13 per cent of internationally important wetlands are likely to be lost over the next 25 years – and the consequences are frightening. As much as 85 per cent of the world's rice is produced in the coastal regions of South and East Asia, and loss of coastal territory in these regions alone threatens the food supply of no fewer than 200 million people. The large deltas of Vietnam, Bangladesh and Burma are particularly vulnerable.

Of course, not all the risks to coastal land are due to rising sea levels. Other human activities such as ground-water extraction (which causes subsidence and salt-water ingress) also result in loss of productive land.

Ports all over the world are an essential part of human life. What effect will rising temperatures and sea levels have on them? New ones are now designed to cope with rises in sea level and even with the freak storms that occur only once in 100 years. In such storms ports are not just put out of action in terms of berthing ships and handling of cargo but also experience severe facility damage. Even so, the frequency of disruption of our existing major ports can be expected to increase, with serious economic effects.

A warmer climate will also affect agriculture – sometimes for the better, but often for the worse. In the United Kingdom, the growing season could increase by about a week – not very much, you might think, but enough to alter the potential for new crops, trees and plants such as maize and sunflower. But then, pests such as the Colorado beetle could become a cause for concern.

Worldwide average rainfall might increase by up to 15 per cent, though with local variations. The United Kingdom may experience more rain, whereas in some parts of the world water resource problems could become seriously aggravated. Pakistan's crop yield could drop by half as a consequence. These variations would greatly affect river flow and ground-water storage, and changes to ocean currents could alter fish populations.

The effects of global warming are well substantiated. It is estimated that a total of 500 million people will suffer hunger over the next 25 years as a result of this threat. It would be catastrophic if we did nothing to alleviate what lies ahead. We must not allow ourselves to be accused of the sin of heedlessness in the face of persuasive, if not totally proven, evidence.

What can and is being done? To guard against global warming, we can start by building up our coastal defences – a horrifically expensive prospect – and by seeking ways of reducing carbon dioxide emissions.

Protecting coastal areas by building defences such as the Thames Barrier in England, creating protective dikes such as is done in the Netherlands, and building quays and docks worldwide to cope with higher sea levels are self-evident, though not necessarily fully effective, measures.

The cost will be phenomenal. The International Panel on Climate Change estimates that $5 billion will be required each year over the next century, on top of the immediate needs of $1,000 billion. The world GNP figure for comparison is $20,000 billion per annum. The distribution of need, however, is very uneven and broadly the poorer countries carry the heaviest burden.

Reduction in carbon dioxide emissions can be tackled in two ways: by improving the efficiency of our present combustion processes; and by converting more of our energy demand to renewable sources and a more fossil-free future. These measures are encouraged by international agreements that aim to reduce carbon dioxide emission levels to 1990 levels by the year 2005.

But what steps are actually being taken? And what do they mean to you? There are, of course, a few energy-saving efforts such as bottle and paper collection, but these are of doubtful validity and trivial in scale. It is not well proven that the collection of used bottles and scrap paper for reprocessing is less expensive in energy and resources than using basic raw materials. Recycling may give people a warm feeling, but it diverts attention from the real scale of the problem. It would be of more benefit – but still of very minor impact – if children walked to school instead of being driven.

Some improvements are being made. More efficient electricity power generating stations are being built, more efficient cars are being produced, more efficient home and industrial insulation is being installed, and all this will undoubtedly reduce carbon dioxide emissions. But this largely happens in the Developed World, which has a population of only a billion out of a global total of 5.75 billion.

If there is to be any chance of keeping the greenhouse gas concentration stable, all possible solutions must be pursued. We must use fossil fuels more efficiently, and fund development of renewable energy sources; remember at present only 13 per cent of primary energy comes from non-fossil fuels.

Significant improvements require international agreement – the air we breathe is universal. Ask yourself, is this likely? Think of the difficulties of the United Nations in getting agreement on almost any subject from whales to Rwanda to Yugoslavia. A second impediment is the time required to change the existing industrial base from fossil fuel extraction and the manufacture of fossil fuel and nuclear power stations to totally new industries developing and exploiting new technologies centred on the manufacture of wave, wind and biomass electrical generation, which are not at all yet well developed. The amount of such industrial change is massive and cannot even be contemplated in decades. Remember it takes 10 years to build one power station of proven technology in an environment with an industrial base which matches these technological needs. The whole idea of conversion of the present pattern of energy supply to one using much lower amounts of fossil fuel requires a time-scale that is so long that population pressures and economic expectations overtake the process of change.

Population control is an important factor. If we are to avoid the futile feeling of walking down an upward-moving escalator, we must do something fast. To achieve a reduction, or even a standstill, in numbers will involve the use of persuasion, and this will depend on sophisticated means of communication. Fortunately, these are available.

The subject of population numbers is generally taboo. Politicians – in democratic countries at least – are not going to ally themselves with such a sensitive subject if it will lose them votes. Anyway, the time-span of parliaments is too short to deal with a problem of this magnitude. And gatherings such as the 1995 UN International Conference on Population and Development are of limited benefit, because conclusions are compromised to meet national and religious sensitivities and few are ever translated into activity on the ground where it matters most.

So the major cause of global warming – the number of humans in the world – is not being tackled with the degree of commitment or urgency that is necessary. Present efforts can be likened to people running around trying to build up the sides of a bath to stop it from overflowing – when really the tap should be turned off.

6 How Fast Are
Our Forests Disappearing?

Deforestation is having an almost incalculable effect on global warming.

Up until the impact of man more than 80 per cent of the world's land was covered by forests. Now the figure is 30 per cent – and falling; a reduction that occurred a long time ago in Europe (*c.* 1000 BC to 1000 AD) but in much of

the rest of the world is happening now. At the beginning of the last century, when the population was approximately 1 billion, the area of forest per living person was the size of 30 football pitches; now it is 3, a tenfold reduction.

Of the 30 per cent, half is in temperate zones and half in the tropical middle band of the globe – this division very roughly corresponds with the developed and under-developed worlds. Broadly speaking, while the West has plundered its original forest acreage over past millennia, the planting of trees is now enabling it to claim sustainable use of its forests.

The logging industry, with its demand for access roads and extraction processes, compacts the soil, which results in rainfall run off because it cannot penetrate the rock-hard ground. The result? Erosion – a feature of all deforestation where sustainable policies are not enforced, extraction methods are not controlled and careful husbanding policies are not implemented.

In tropical regions, the picture is very unhappy. Around 1 per cent of the tropical forests are lost every year – which means that on these trends, tropical forests will disappear within a century. With them will go their 'lung' capacity, not to mention the incredible array of natural life that they contain. If 1 per cent does not sound much, consider this: it is the equivalent to an area the size of Britain. It is a loss of 80 acres, roughly 80 football pitches, every minute of every day of every year. Now you see the problem.

Most deforestation is due to clearance for agriculture, ranching or animal husbandry. It is this land hunger that drives deforestation, together with the loss due to improperly conducted logging operations. Clearance for industry and urbanization is locally important but not a big factor in the global equation.

The land taken for agriculture is the result of pioneering efforts as man moves ever outward from his village or town to seek a more secure and sustainable standard of living. This short-sighted expansion is encouraged by governments, which provide subsidies and tax breaks. Deforestation actually gives a double benefit to the State: it helps solve some of the space limitations brought about by rising populations, and there is potential income from the sale of the felled timber, though all too often the timber is burnt on site.

While it could be piously argued that government incentives should be discouraged, it is clear that local communities are entitled to exploit such economic opportunities. The pot would be calling the kettle black if we did not recognize that the West has, in the past, massively plundered its own forest resources. And as we consume our precious coal and oil reserves, we continue to do so – albeit indirectly.

There is a popular misconception that most forest land converted for agricultural use is to meet the West's (and particularly North America's) demand for high protein meat products such as beefburgers. While our requirement is indeed substantial, demand is even bigger in those very countries whose trees have been cut down – such as those in South America. In fact, the demand for meat products in general is greater in Latin America than it is in North America; and the population, of course, is increasing.

One of the reasons for cutting down forests is to meet the needs of expanding cities. As people seek employment, a home of their own and a means of transport, so they move away from their villages to the towns. But then what happens? The towns themselves expand – perhaps even into the very villages from which those people have come in the first place. In the West there have been attempts to contain urbanization by 'green belt' policies – but not in the Underdeveloped World. The slums surrounding the mega-cities of Buenos Aires (13 million). São Paulo (17 million), Mexico City (19 million) and Calcutta (11 million) have become greater in area and population than the core cities themselves. With such gigantism come all the horrors of a breakdown in law and order, impossible housing problems, unattainable job creation prospects and so on; things so serious that they threaten the very stability of nations.

We all seek higher living standards – of course we do. But have we in the West the courage to say, on the grounds of reducing deforestation, that the 4 billion under-developed of the world should not experience economic growth? No elected politician would propose such a thing. And no dictator would put at risk the stability of his regime if he did not keep alive expectations of improving economic conditions.

While planners are predicting 3 per cent economic growth per year, a doubling in 25 years, the actual growth per capita is reduced to only 1 per cent if we take population growth into account. The one policy that could really help solve this dilemma is to seek to reduce numbers. The average number of children per family in tropical regions is of the order of four. This must be reduced to below two. The time to achieve anything along these lines is horrifyingly short. Think of Buenos Aires, São Paulo, Mexico City and Calcutta all becoming ever bigger in size and growing beyond man's ability to organize.

7 Is There Enough Food?

Food is a basic human necessity – yet millions of people in the Underdeveloped World are malnourished and many more hover on the brink of starvation. It is among these people that population growth is most rapid. So what are the prospects for Africa in 25 years time when the population will have doubled?

Most food consumed by the world's population is in the form of cereals – wheat, maize and barley – and grasses such as rice. It is possible and convenient to measure all food production and consumption on the common basis of nutritional value. For example, a grilled steak has the same cereal equivalent as two large bowls of boiled rice, a large tin of baked beans or a loaf of bread. Another way of expressing nutritional value is in calories. In principle this is no different from the way we compare the energy content of different fuels such as coal, oil or gas.

So how much food does a person need to take in each day – both in terms of the minimum for survival and the minimum for reasonable health and an active life? The survival diet is 1,600 calories per person per day. That required for health and activity is 2,500. The average calorie intake in the world today is 2,600 – around 3,500 in the West and 2,400 in the Underdeveloped World – of whom half actually get less than 2,000. Worldwide, the range of intake is very large: the average Belgian consumes about 4,000 calories compared with the Ethiopian's 1,700 – a ratio of more than 2:1.

Three thousand five hundred calories a day is equivalent to a ton of cereal a year. The total annual consumption of the billion people in the West is, therefore, in the order of a billion tons, while the 4 billion people in the Underdeveloped World consume just under 3 billion tons. The Developed World is exceeding the minimum required for health and activity by about one third – and many of our modern diseases arise from this fact, heart disorder being but one example. The Underdeveloped World as a whole is just short of the desirable minimum; but this is an average. One in twenty is at or below the bottom of the survival scale.

Can the world feed its present population? Is there enough land? Is there enough water? And is there enough political will?

There is enough rainfall in the world to grow all the food necessary, but it is so unevenly distributed that water itself becomes a limiting factor in meeting the demand. (See Chapter 8)

There is also no shortage of land. Even in densely populated countries such as India the amount of land unused or under-used is considerable. But the *suitability* of land for agricultural use is a different matter. A third of Africa's land mass consists of the Sahara Desert – so couldn't it be irrigated? The answer is yes – but the problems are finding a reasonable adjacent water source and the cost of delivery.

If the Mediterranean was to be the Sahara's source of water, desalination would be required to make it potable. This could be achieved only by distillation – a very expensive process in energy terms; and the pumping costs would be astronomical. For the poor countries of the Underdeveloped World, the economics are quite unattainable.

In the Underdeveloped World, fertile land for crop growth has diminished due to desertification and to population increase from two-thirds of one acre each (an acre is about one football pitch in size) in 1970 to half an acre in the 1980s; a 30 per cent decrease. Shortage of usable land is reaching a critical point in many areas: the people of West Africa are already using all their potential farmland, and the population is expected to double in the next 25 years.

Soil exhaustion and drought also cause vast losses of fertile land. It is estimated that about one third of China's productive agricultural capacity has been lost in this way in the last forty years.

Can the 1 billion inhabitants of the Developed World produce enough food to bring the 250 million severely malnourished inhabitants of the Underdeveloped World up to the minimum required for a healthy, active life? Yes! The 60 million tons required is already produced and distributed every year – but it is done unevenly. Realistically we cannot expect those 250 million to feed themselves. To bring their standards up to that required for a healthy and active life requires an increase in their own present food production of about 50 per cent and that is an immediate requirement; overnight so to speak.

There is too much food in the West – so much so that we pay farmers to set aside their land in an attempt to cut down on over-production! This practice was introduced to prevent the build-up of 'mountains' of meat and cereals and 'lakes' of wine.

On the other hand the Third World has barely been managing to keep pace with its own growth in population over recent decades. There is not the remotest likelihood of their higher food production ambitions being met, given that the population of the Third World will increase by over 50 per cent in the next 25 years; it is difficult to be hopeful.

Total global food production is presently roughly static; slightly decreasing in fact. In recent decades total output has increased in line with population growth and therefore the amount of food produced per person has remained the same. The seriously malnourished though have known no mercy, experienced no respite.

As already noted one of the underlying causes of famine in parts of the world is inadequate rainfall, which has a direct bearing on land productivity. The other is land quality. So could huge amounts of fertilizer help? In the West, fertilizer consumption per individual is 70 kilograms compared to the Underdeveloped World's 15. To match the West, the production of fertilizer would need to triple. During the many years it takes to build even one fertilizer complex, the population time bomb continues to tick at an alarming rate.

But fertilizer alone is not the answer. The output of the soil simply cannot be increased infinitely by adding more and more fertilizer. For all sorts of reasons the law of diminishing returns applies: the gain from extra fertilizer gradually gives less and less reward.

Food is only food when it is on the table in front of you and can be eaten. A capability to grow food does not satisfy hunger. Many people disguise the unease in their mind about the problem of world food supply by saying to themselves that there is plenty of land available; they do not take into account the problems of water, land quality or fertilizer availability. They are conditioned to hearing about the surpluses of the developed nations but do not ask why these surpluses are not distributed to the hungry of the world. As has been shown earlier in this chapter there is enough excess to do that at present although maybe not in the future but the actual growing of the food is just one of the problems and not the biggest one at that. Other difficulties include those of distribution in the technical sense of physically transporting the food and another is finding the means of paying for the growing of the food. Farmers do need to be paid. Yes, the money spent on practices such as 'set aside' could be used to pay for the donation of food but it is a relatively trivial amount. Any increase in taxes to pay for the growing and sending of food to, what is perceived to be, some far-off land on a large and permanent basis is just not acceptable to the tax payer. Such generosity on the required scale has never been demonstrated; we choose 'set aside' rather than distribution. Maybe it should not be so but it is.

Heartrending scenes of starvation in Africa shown on television do stir our consciences but the impact quickly fades. People like to talk about their generosity and willingness to help in worldwide concerns like this but as any politician will tell you when the problem comes anywhere near their own individual pocket or to the imposition of more taxes their vote is cast in another direction. Keep in mind that when we talk of 'they' we really mean 'we'. This human response of self-interest should not be condemned; it is important we do not hide behind the view that fundamental goodness will eventually solve the problems. There is no intention on the part of the author to denigrate humans for acting self-interestedly rather than on goodness and generosity principles; or vice versa. The intention is to focus solutions on what is required taking cognizance of what actually happens and not what should happen.

At the end of such a chapter as this it is difficult for us to remain optimistic. One can only hope that as the world becomes smaller day by day, as transnational and international institutions develop and as technology advances in fields such as genetics some relief to the problem of world hunger will emerge but the figures are not at all reassuring. To await a second green revolution is dangerous; and somewhat futile in that the benefits of the first were dissipated within a few years by the growth in population.

While we do not have any ready solutions at least we can do what we can to avoid seriously aggravating an already desperate situation. It behoves all of us to take whatever steps we can to prevent any further growth in human numbers and, in fact, we must aim for a falling trend. This is of course the core subject of this book. The urgency of the matter should be kept in the forefront of our minds by the reminder that the population of the world will go up by 50 per cent over the next twenty-five years which is in your lifetime if you are under 50.

8 Is There Enough Water?

Water is fundamental to all life on earth. Without it, the planet would be dead. But is there sufficient to meet the needs of a burgeoning population? How well is it distributed and how practical would it be to redistribute water from a plentiful area to an arid region?

All drinkable, available and usable water comes from rain. Tiny amounts are produced by distillation of sea water in Kuwait, Saudi Arabia and a few other places. Other potential sources are the ice-caps at the North and

South Poles – but these are not really available to us in practical terms, witness the fairly unsuccessful attempt to tow icebergs to regions where water is needed. Underground water is tapped in many areas, but these reserves are replenished by rain, and are not an infinite supply. Rather, such aquifers act like a big sponge soaking up any surplus rainfall and holding it for when drought conditions come along.

Some 10 per cent of the world's collected water is consumed domestically, 70 per cent is used for agriculture, and the remaining 20 per cent by industry. In the more arid areas of the world domestic use accounts for virtually all collected water.

Oceans cover two-thirds of the earth's surface, but the irony is that salt water is not suitable for industrial, commercial, agricultural or domestic purposes.

The world's mean average annual rainfall on land is around 12 inches – the equivalent of 4,000 gallons per person per day. The minimum amount required to feed someone for a day, including that used by agriculture to grow the food, is 200 gallons per day. So there is more than enough rainfall to supply all non-commercial needs everywhere. Unhappily, however, rainfall is not evenly distributed, either geographically or seasonally. Think of Far Eastern monsoons, where up to 30 inches can fall in an hour, compared with a light shower in Europe!

Today about 2 billion people – a third of the world's population – in 80 countries live with chronic water shortage and the situation is worsening by 200 million per year.

To take one example, the Nile Valley, which embraces Egypt, Sudan, Uganda and Ethiopia, has an annual rainfall of 8 inches. But as much as 80 per cent is lost by evaporation, and to maintain river-flow 10 per cent necessarily flows into the Mediterranean. In other words, only 10 per cent of the Nile Valley's rainfall is available to be harnessed for human use. At 200 gallons per day this is only enough for 250 million people; only 80 million allowing for industrial uses. The actual population is 130 million and rising at 4 million per year. Water demand will double within 25 years.

But what about the Aswan Dam? The problem is that it does not *produce* water; it merely collects and distributes it as and when required. Certainly it reduces the loss of useful rain-water in the sense that without it any uncollected excess would run via the Nile into the Mediterranean. Already its benefits have been outstripped by demand. When the dam became operational in 1970, it was expected to produce enough water to meet the needs of 10 million people – but the population of Egypt is increasing by some 2 million a year.

This contrasts starkly with Europe and some Far East countries where there is abundant rainfall and enough economic muscle to invest in many dams and reservoirs, but even in these areas, water shortages occur. In Britain during the summer months shortages are reaching such a level that water meters are being considered as well as recycling.

The collection of water from rainfall is very efficient in some areas of the world. Extraction from the Colorado River, which drains about a quarter of the land area of America, is so high that in some years the river flow never reaches the sea. But even America has its water problems; think of the dust bowl of the Middle West.

Probably the world's most serious shortage occurs in the North China Plain, a semi-arid region with a population of 200 million and including major cities such as Beijing. Within 5 years there will be 10 per cent less water than is necessary and Beijing alone will have a shortfall several times that figure.

Generally throughout the world desert areas are increasing at an annual rate equal in area to that of Great Britain and it is expected that this trend will continue.

The disparity between countries with adequate rainfall and financial muscle and those without is expected to widen, in part, because of the predicted effects of global warming. So what can be done? How are extra mouths to be fed and watered, particularly in arid areas where the numbers increase each year by 15 million? Pumping of sea water over very long distances is beyond the means of most countries as is the straight economic cost of distillation: Kuwait produces 3 per cent of its water needs by distillation and the capital cost of the plant is astronomical – even for a country with such vast oil and gas reserves.

Another option is to pump naturally drinkable water from one area to another. In Ethiopia there is plenty of rainfall in the mountains but serious scarcity in the Tigrei and Wollo desert regions. However, this too, is prohibitively expensive. To pump enough water for, say, 10 million people over 500 miles, the distances involved in this example, requires the equivalent of 10 large power stations. The total cost of power, capital and revenue on a per person basis just to meet domestic and food growing needs would be in excess of $2,000 per person per annum. Compare this with the average income of the Under-developed World of $500 a year.

To say the unsayable, some people are in the wrong place. Is it beyond the wit of man to think of ways to relocate them to areas that are less prone to water shortage? The difficulties are, of course, immense: quite apart from transportation, there is the whole question of land ownership and rehousing, not to mention nationality, political and trans-migration issues. But if we do not come up with something more radical than the present *ad hoc* emergency aid approach, the distressing pictures that we see on our television screens each year are likely to continue and become even more appalling.

9   What About Pollution?

Only one third of the earth's surface is land of which about a third is covered by forests. Another third comprises arable and pasture lands, and the remaining third is either unused or unusable – deserts, for example.

Usable land is exposed to a variety of environmental hazards. Soil degradation polluted by the burning of fossil fuels is one, heavy metals in the air another, and over-intensive use of agrochemicals such as pesticides is yet another. These pollutants affect lakes and ground-water, too.

Degradation also occurs because of salt deposition caused by heavy irrigation in rural areas, and by abandoned industrial sites in urban areas. Add to this our demands for agriculture, forestry, industry and energy expansion, for housing, mining and tourism, and it becomes obvious that there is a diminishing amount of usable land available to meet the needs of an ever burgeoning population.

All this puts great pressure on the productivity of the soil. Inappropriate practices, deforestation and poor husbandry result in taking out more from the soil than it can sustain – something that inevitably leads to erosion. In about a third of the World's arable crop lands the soil fertility is in decline because of agricultural over-cropping. To a significant extent, industrialized countries can compensate with increased use of fertilizers, but in the Third World soil fertility loss is reducing the land productivity with no chance of reinstatement. In China, degradation of arable land has increased to such an extent that desert regions now exceed fertile areas. World-wide, an area approximately the size of Britain becomes desert each year.

It is hardly any wonder, therefore, that the world is seeking to expand its agriculturally usable surface area by irrigation schemes. But in arid and semi-arid regions, the high evaporation rate brought about by tropical temperatures is causing salts from irrigated water to be left on the fields. Over time, this saline accumulation damages plants and the soil becomes unproductive. A measure of this can even be seen in the United States. Of the arable cultivated land there, approximately 10 per cent is irrigated – and as much as a quarter of this has been damaged by excess salts.

The heavy and ever-increasing use of fossil fuels results in discharges into the atmosphere of very large amounts of sulphur pollution, which travels long distances through the

atmosphere. Some of this is deposited on land and some in lakes and rivers, causing serious fertility problems.

Acid rain, the end product of sulphur emissions, is one of the major industrial blights to hit the world in recent times. Like all forms of pollution, it knows no boundaries, respects no frontiers.

Acid rain is caused by gases from smoke and exhaust fumes mixing with water. This creates dilute acid, which falls as acid rain. By far the biggest culprit is the West – but it is also the victim of its own polluting practices: estimates suggest that half of Germany's forests have been destroyed by acid rain, and in the United Kingdom as much as two thirds of forests are affected by this blight. In Europe, an area roughly the size of Scotland has been destroyed by acid rain over the past century. Current research qualifies these concerns about acid rain and more stress is being focused on the harmful effects of ozone from motor car exhausts. Experts say that the level of European sustainable timber production is reduced by a fifth because of harmful emissions.

As man demands higher and higher standards of living, more energy, more coal and more electricity is demanded. The by-product is acid rain and other atmospheric pollutants. It would be a brave politician who would advocate resisting the pressure for higher standards of living in order to reduce this hazard. There is, fortunately, an alternative – concentrating efforts on reducing population numbers and, thereby, the adverse pressure on the environment.

Atmospheric pollution is but one facet of the problem. In most countries, the exact quantities of waste from industry are still unknown. The West produced about 5 billion tons a year in the early 1980s; now, we are perhaps 50 per cent up on that. The Underdeveloped and Developing World produce two thirds less than the West – and that for a population of 4 billion compared to the West's 1 billion! If the world doubles its economic activity over the next 25 years, it will generate at least twice the present volume. Incineration, landfill and dumping at sea are the main methods used today. But the above-ground sites are becoming less and less available and the costs of disposal are inescapably rising.

This takes no account of the problem of hazardous waste, of which there is about half a billion tons a year. At least 50 per cent is chemical-based, and you do not need to be an avid listener to news bulletins to be aware of ships carrying chemicals and radio-active waste around the world seeking an asylum country that will allow them to discharge their unsavoury cargo. Virtually all of the Western World exports some of its waste, with Germany being by far the biggest culprit.

Touching briefly on municipal or domestic waste, indicative figures are available only for the Western World. The domestic waste per capita in the West was around 900 lb in 1980 and it is now about 1100 lb, a 25 per cent increase. These averages mask substantial differences between countries: for example, United Kingdom produces 700 lb against 2000 lb in the United States of America. Think about it; we generate something like 10 times our own body weight each year in domestic waste. In the West the problem is the disposal of this vast amount; landfill sites are becoming very scarce, and those available and usable are becoming increasingly remote. We have hardly seen anything yet.

Many countries regard oceans as the ultimate sink. Some wastes are discharged directly by vessels and rigs at sea, and others are transported from distant reaches by rivers and by the atmosphere. Land-based sources are responsible for the bulk of wastes entering the oceans; they are chiefly in the form of sewage, industrial discharges and agricultural run-off. Waste inputs include surplus nutrients from agriculture, oil discharges, radio-active wastes, plastics and trace heavy metals including mercury, lead and cadmium. Many of these contaminants are eventually ingested by fish and, once in the food chain, become a direct threat to humans.

Much waste originating from such land based sources is deposited in the fertile coastal zones, often in or near sensitive productive marine areas. A distinction, therefore, must be made between the pollution effects in the coastal zones, which are highly sensitive and stressed regions receiving the concentrated waste inputs, and the open ocean, a vast area receiving proportionately fewer wastes with more diffuse sources.

Several billion people in the world rely on fish for their main source of animal protein. The ever-expanding population demands more fish, and already the oceans are stretched to breaking point. The catch of fish and crustaceans has increased from 50 million tons in 1965 to 100 million tons today. A further 10 million tons can be added from fish farming. It is estimated that we have reached the maximum global sustainable yield of 100 million tons, excluding aquaculture.

Fishermen are now seeking out more exotic species to take the place of exhausted traditional staples such as cod and haddock in the North Sea, and there is increasing pressure to catch smaller and younger fish, thereby further depleting breeding stocks. More and more sophisticated fishing boats are being built and some are so efficient that they need operate for only a few weeks a year before internationally agreed quotas are exhausted. Half-hearted attempts have been made by western nations to decommission fishing vessels, but in world terms the trawler fleet has grown: the number of vessels has doubled in the last 20 years.

It is not to the sea, therefore, that we can turn to feed the extra mouths that will be born over the next 25 years. It can be hoped that aquaculture can make up some of the difference but on a proportionate basis with a 50 per cent increase in the world's population in the next 25 years one would need something like a 5 times expansion factor in the field of fish farming. This requires massive capital expenditure and suitable sites, neither of which are going to be readily available.

The concern of this book is not to propose solutions to these problems, a task that is well beyond the capability of the author. In any event, there are experts tackling these many difficult areas. The problems generally are not of an insoluble technical nature but rather are concerned with the speed at which pollution and contamination grows as economic activity and people numbers increase. Unlike the problem of expanding the industrial base, which can at least be decided by each country in its own interest, many of the marine environment problems are international and getting agreement amongst the countries is an exceedingly slow process. For example, think of the time and effort devoted to trying to get agreement on whale catching quotas.

Every attempt should be made, of course, to reduce the pollution of the oceans in order to increase fish stocks, and to police the fishing industry to ensure that quotas are adhered to and the minimum sizes of net are not exceeded. All that is self-evident, but equally self-evident must be the desire to take some pressure off the environment by reducing the increase in population numbers.

10   The Impact of the Motor Vehicle

There are 700 million vehicles in the world today, of which three quarters are in the Developed World. There are 450 million cars, 110 million light duty vehicles, 110 million motor cycles and 30 million heavy duty vehicles and it is projected that by the year 2025 the number will more than double. They consume one third of the world's oil production and add massively to atmospheric pollution. Some 65 million vehicles are produced each year.

The biggest increase will happen in the Underdeveloped World. But even in the West the increase will be not far short of 100 per cent. Think of the frustration caused by today's traffic conditions and what *any* increase in vehicle numbers will mean *to you*.

The costs of congestion are substantial, both in terms of vehicle costs and the wear and tear on human beings. Traffic snarl-ups increase fuel bills by 5 per cent or so, and maintenance costs by 10 per cent.

Annual average travel distance in the West has increased by 50 per cent over the past two decades. The wear and tear on human beings is worrying. Already traffic queues at weekends and holiday periods in the West can be up to 50 miles long and generate massive frustration and road rage. The average American spends 1,600 hours a year in his car – that's equivalent to 67 days and nights – yet he travels only 7,500 miles (less than 5 miles per hour). One has to

ask … is it really worthwhile? Is there not an alternative?

That's one question. The other is whether the vehicle problem can be solved. In Britain, the Royal Commission on Environmental Pollution urged a radical rethink of transport policies. It recommended an increase in the proportion of public transport journeys from 10 per cent to 30 per cent within the next 25 years. But even if this were to be achieved, the total number of journeys would still increase by 10 per cent, due to the increase in vehicle numbers.

The Commission also recommends stopping road building and increasing fuel duties. Both policies are intended to discourage the private use of vehicles – but the consequences would be serious for the West in respect of GNP and employment, and therefore politically unpopular: vehicle production and road building account for around 5 per cent of GNP and 4 per cent of jobs.

Few, if any, reviews concerning the future of road conditions discuss people numbers. This is avoided partly because generally population control is a taboo subject and also because most people, but politicians in particular with votes on the line, realize that the ambition of most of us is to have our own private vehicle. The proposals to reduce the impact of growth in vehicle numbers by increasing fuel duties and reducing the road building programme have merit. But of equal merit is a policy aimed at trying to persuade the policy makers, and ourselves, that much of this projected doubling of vehicle numbers over the next 25 years arises due to the ownership aspirations of the predicted 50 per cent increase in the population. A bigger impact, by far, would be achieved if we endeavoured to curtail people numbers rather than seek to reduce car ownership ambitions (the thought of even more traffic congestion in Buenos Aires and Calcutta is almost stupefying!)

It is folly not to see the growth in population as being one of the major factors in future transport projections. Clearly it is one of the spokes that buttress in place the hub of the wheel, which says, 'We must seek to reduce population numbers'.

11 Unemployment

Unemployment is the one common area of concern about which most societies agree. Control of population numbers can realistically contribute to solving this problem.

All countries endeavour to create full employment. The West is losing out on two counts – the unstoppable flow of jobs to the Pacific Rim countries, and others, with a lower standard of living, and the rise in human numbers.

The international drift of industry and commerce is now beyond the control of any one government, and to hope that the West's technological sophistication can halt it is forlorn. In today's world, countries such as Korea and China are every bit as technically sophisticated and aware as the West, but have labour rates of pay up to 20 or 30 times lower.

The promises of political parties, particularly when they are out of power, that they will remove the scourge of unemployment by job creation should not only be taken with a pinch of salt; they should not be countenanced at all. In an international world where free economic and technological interchange exists, politicians have negligible control over trading policies anywhere. No nation can escape from this jungle of international complexity. Also job creation with its inescapable energy requirement has its inevitable consequence on global warming.

The population in the West is rising by 5 million per year taking out confusing factors such as immigration and emigration. Europe, for instance, experiences an increase

in population per annum of 700,000 and it is 2 million in
the United States of America. The level of unemployment
in Europe, for instance, is presently 15 million and is
expected to grow as the centre of gravity of commercial
and industrial development moves to the Pacific Rim and
automation of clerical and manual work continues to race
ahead. If we do nothing, that unemployment figure will
massively increase and with it the likelihood of your child
becoming an unemployment statistic.

Reducing unemployment in a rising population is rather
like trying to run down an upward-moving escalator when
it is outpacing you. If all population growth could be
stopped overnight in, say, the European Union the number
of people seeking work would be 700,000 fewer each year
15 years from now.

Some people may argue that in the past, rising
population numbers have been associated with higher
economic activity and a rising standard of living – so
falling numbers could therefore have the opposite effect.
There is, however, now no correlation between numbers
employed and economic well-being. Technology has been
uncoupling this relationship, if there ever was one, over the
past two decades, and will continue to do so apace.

It could also be argued, loud and clear, that a child born
today is not an unemployment statistic for at least 15 years
and, in any event, it is not possible to stop population
growth overnight. Absolutely true. But a population
control policy should not be rejected because of the long
gestation time. If Margaret Thatcher had started a
campaign to control population growth in 1979, when she
came to power, we would now be reaping the benefits of
that policy. But she did not, and I really wonder why this
approach to reducing unemployment has not been taken
up. Politician's prefer to make vague assertions about
creating more jobs without saying how they would do it. I
can only suppose that they feel that the votes to be lost by

advocating population control exceed those to be gained.

This inhibition is also true of 'green' charities. They have often found through bitter experience that every time they take a stance on population control, the level of financial support they receive is seriously reduced. We could help reverse this by supporting only those charities which specify that some or all of their donations are devoted to birth control measures.

In both the Developed and Underdeveloped World, there is a growing awareness of a need to do something to stop the inexorable rise in numbers. We can expect the politicians of all hues and from all nations soon to become convinced that the time is now ripe to take on board the theme of population control. The balance of political advantage is there to be grasped now – not later.

12　Animals and Other Matters

In a book of this size it is not possible to cover, in detail, all relevant topics that contribute to my main argument. However, it is true to say that the origin of most of our anxieties – pollution is one example – can be attributed to a rise in living standards and to the large increase in the world's population.

What does the future hold? With predictions of rising standards of living and a 50 per cent increase in people numbers over the next 25 years, we cannot avoid the logic that says the fewer people there are, the less impact there will be on the environment.

Meanwhile, we have to feed the world. Although intensive farming is a practice about which many people feel uncomfortable, we do not know any other way of meeting the food needs of a rising population. Battery chickens whose legs have atrophied, and pigs and calves confined in spaces so small that they cannot even turn round, are but two examples of our disquiet.

Such treatment of animals is not new: geese have been forcibly fed for 100 years or more to produce *pâté de foie gras*. The whole process of animal farming has become one of protein production, with animal husbandry taking second place. These days we are even trying to produce featherless chickens because feathers are regarded as a nuisance!

Little can be done about the clamour for higher standards of living, nor would one want to as all of us have aspirations for a better lifestyle. What we can do is alleviate the situation by taking active steps to prevent an increase in human numbers.

There is, of course, a difference between what we say and what we do. Whilst many people feel uncomfortable about intensive farming methods in theory, in practice things are often different. In a marketing study, 80 per cent of respondents said they would prefer to buy free-range eggs rather than those produced by intensive farming methods – but only 10 per cent were prepared to pay a marginally higher price for them. Money speaks louder than words.

Another area of concern is the fall in biodiversity – the huge range of species in the world – and the decline in total numbers, particularly in the animal kingdom. Think of the vast numbers of buffalo that once roamed the American continent – 60 million in 1880, now down to under 100,000; and most killed in a brief 10-year period. Reflect on the efforts being made to ensure that tigers, pandas,

whales, koalas and elephants may live. Think of the policing required to stop the imports and exports of ivory. Consider the threat to the rhinoceros because its horn (which fetches more than the price of gold on the black market) is used in traditional Oriental medicinal practices.

Tigers, pandas, whales, koalas and elephants are known as 'charismatic' animals for exactly that reason. They get all the attention. But there are many other species – fish, insects, birds, even mosses and lichens – that are also under serious threat. The lapwing in Britain, for example, and the cod and haddock of the Northern seas. Our quandary is that for as long as humans need food, it is difficult to ask them to preserve an edible species, particularly when we ourselves are sitting at a rich man's table.

There are anything between 5 and 30 million species in the world, with invertebrates forming the largest group and mammals the smallest. Of the million species in the West it is judged that 40 per cent of mammals, 10 per cent of birds and 50 per cent of fish are threatened.

Yet our own numbers continue to rise. Crowded holiday beaches are an irritation, and the snarl-ups on the roads are a major annoyance. Away from the madding crowd, access to mountains and remote regions is now being restricted because of the ecological damage caused by those very people who seek the escape offered by uncluttered space. The path to the top of Everest has become a littered highway. Freedom of access is causing such damage that more and more 'Keep to the Path' signs are being erected; it will not be long before the message is 'Keep Away'.

Another concern is man's ability to organize and control himself. Throughout the world the bureaucratic and legal processes are grinding to a halt because of scale and complexity. This is an inescapable consequence of commerce and trade becoming more international and the creation of more and more political federal structures of one form or another, such as the European Union. The complexity of these legal and administrative systems increases massively and disproportionately with size; chaos theory.

The numbers of people, *per se*, is another cause of paralysis of such systems. Clearly the more people there are the more the legal and administrative processes will be on demand. To be silly about it, if you had only two families in the Western World it is unlikely that one would have to have any legal or administrative systems to handle their relationship problems. Problems arise at the interface between people, organizations etc. More and more statutes are required to meet the increasing sophistication and refinement of people's interaction with one another; think of labour laws, laws of inheritance, laws of the sea, tax laws, divorce laws and so on.

The fewer people there are the fewer will be the boundary interactions between them with the risk of disagreement. Unfortunately the inescapable mathematics is that the number of such boundary interactions does not go up in direct proportion to the number of people but at a faster, exponential rate; between two people there is but one, between three people there are three, between four people there are six, between five people there are ten and so on. So the pressure created on our legal and administrative systems is, in part, a result of increasing refinement and sophistication but also, because of the increase in numbers resulting in disproportionate increase in the number of possible disputes. The increase in the world population over the next 25 years must therefore pose a very serious threat to the stability of our legal and administrative systems.

Rising crime rates are clearly a matter of serious disquiet both in the Developed and Underdeveloped Worlds. The causes are manifold and include the drug culture, unemployment, loss of accepted social norms, the envy of poverty and sheer numbers. The impact of this last point is best illustrated by reflecting that the real dramatic rise in crime rates has been in the urban rather than the rural areas.

13   The Ground Rules

By now I hope you are convinced that the rate of growth in human numbers is a matter of grave concern. Whilst it would be unwise to declare dogmatically that the evidence contains no uncertainty, I am firm in my view that it would be imprudent and shortsighted not to take strong precautionary action now.

There are, of course, no easy options. That's the first rule of life. Whatever may be proposed is likely to cause pain somewhere. If, for example, the West decided to insist on a high percentage of its aid being spent on contraceptive measures, fewer people in the receiving countries would be freed from hunger – at least in the short term. Even if the amount of aid is increased, and there is no evidence the Developed World is willing to do this, the decision still has to be made as to whether the extra aid should go towards feeding that hungry child so poignantly shown on television or insist that it be spent on contraceptive measures.

Unfortunately, difficult decisions are required. There are no painless ways out. Ignoring the problem will not make it go away. The greatest evil would be to do nothing.

The second rule is that we live in a world of uncertainty and probability. Even tomorrow's dawn is uncertain. There are some massive chunks of rock out in space any one of which could hit us like the one that hit Jupiter recently. If one of these hit us, incidentally it does not need to be a big piece, a cubic mile would do, it would wipe out the human race. Fortunately the probability of this happening is very low. Scientists think in terms of levels of probability rather than certainty. They do not talk of the certainty of fingerprinting, but claim there is only a one in a hundred million chance that any two fingerprints are the same. In considering the evidence for claiming there is an overpopulation problem, we, too, must live in this same world of uncertainty and probability; while the evidence is exceedingly strong it is not indisputable. The same goes for any proposed solutions.

The purpose of the metaphor used in the preface of this book – the hub of overpopulation buttressed in place by the many spokes of its appalling consequences – is to persuade the reader to take all factors into account before reaching a judgement. It is useful to extend the metaphor by considering that each spoke, if taken separately, has its own degree of uncertainty – but taken together, the level of confidence in the conclusion becomes much greater.

The third rule concerns the time horizon against which decisions are made. Are we doing good in the short term if more misery is created in the long term? Short-term aid to countries that currently cannot feed themselves may in the end cause more misery because the children of the next generation will experience even more deprivation. The reverse is where action, which has dramatically unpleasant effects in the short term, is taken in order to alleviate a future horror. Present Chinese efforts to restrict population growth are undoubtedly causing pain and even misery in the short term, but in their Government's view the horrors they would experience in the longer term would be even worse. So we need to find a balance between short- and longer-term interests.

The fourth rule is that self-interest will dominate and altruism will not.

It seems to be a futile exercise to try to persuade those in the affluent areas of the world to sacrifice their aim of higher living standards and a conspicuous consumption of resources in order to lessen some longer term concern about human survival. We all find difficulty in taking seriously an anxiety which is far-off in terms of time or geography. But the dependence of mankind on our small globe, on space-ship earth, is such that there is *no* far-off country.

Those in the West who want a car – often a second car – and who wonder whether to have a country cottage as well, who feel they need a holiday, who are not as well-off as their immediate neighbours, do not compare themselves with someone in the Underdeveloped World who has an income of less than $500 a year. We all make comparisons with our immediate neighbours and often our desires are driven out of envy.

Entreaties to the goodness of people to help alleviate starvation, to take massive and long-term action to lessen threats to life on the planet fall on deaf ears. Appeals by institutions from the religious to the secular seem to be of little avail except at the margin. Solutions to the world's problems are not likely to come from appeals to goodwill but rather by persuasion that what we do is in *our* best interest. Ask any politician and he will tell you that a proposal to put up taxes to pay for the good deeds loses votes; only minor sacrifices at the margin can be asked for – witness the famous phrase by an American president, George Bush, which it is claimed won him the election, 'Read my lips, no new taxes!' Treat with suspicion evangelical political speeches, particularly those containing pious platitudes. People will only reduce their family size when it is in their interest to do so. Success in reducing population numbers and growth will only come after persuading people that such a policy is in their own interest.

The fifth rule says there is absolutely no point in the West preaching to Underdeveloped countries to restrict their population if we are not willing to make our own contribution. It is important politically that the West is seen to contribute and not put all the pressures for action on the Third World.

The disparity between the consumption rates of the Developed and the Underdeveloped areas of the world is some 30 to 1 on a per person basis and some 8 to 1 on a total consumption basis of 1 billion to 4 billion inhabitants. In a perverse way it could be argued that to maintain a spirit of equity the Third World could justify having 8 or even 30 children to the West's 1.

People of the Underdeveloped World resent being criticized for plundering their great forests when we readily plunder our own resources of oil, coal and other materials to attain ever higher standards of living. We have the impertinence to ask them to stop cutting down one of their only great economic resources so that we can discharge large amounts of carbon dioxide, a consequence of our high economic activity, and rely on the lung of the great far-away forests to absorb at least some of it!

There is a lot the West can do, by way of example, if we want to show the world that we are treating the problem of the population bomb seriously. Our total numbers are still increasing in the West even though our women are now having fewer than two children each. The increase in total numbers is due to the present high proportion of women of child-bearing age in the population and, without intervention, it will take too long a time, some fifteen years, before our population stabilizes; one of the important themes of this book is that time is of the essence. Clearly we should aim for even lower reproduction rates not just on the grounds of total numbers but also to demonstrate to the rest of the world our identity with the problem.

We have the most sophisticated means of communication, which enables us to enter every home; we have the most sophisticated medical services, which can deliver the best technical birth-control advice and the manufacturing capability of producing contraceptive devices. Think of the great propaganda exercise undertaken over recent years to stop people smoking! It was impressively successful – so why shouldn't we try the same for the population problem? Can we not persuade those who are dedicated and single-minded enough to fight the war on smoking to put their energies into the hugely more important campaign to control population growth?

I acknowledge that there are the problems of religious and nationalistic beliefs to overcome, but there is a mood abroad that the time is ripe for a successful attack on the subject. Efforts asked of the Underdeveloped World must be matched by equivalent sacrifices in the West.

14 What Is To Be Done?

It seems abundantly clear that if we continue to ignore the population problem, the outcome could be cataclysmic for mankind – because if we leave things too long, it will simply be too late to rectify the consequences. In the animal kingdom, of which we are part, when populations reach resource limitations they experience a crash in numbers with the attendant horrors inescapably linked to such a fall. Better to take prudent steps now and be proven wrong than to take none and find out too late that we should have taken action earlier.

What can we in the Western World do to help resolve the threat? Very little if you are a pessimist – and not much more if you are an optimist! But the consequences are so dire that we must take corrective action now.

By making more efficient use of our consumption of energy and material resources, the West can contribute more to reducing the threats to civilization than the Third World can. In the Underdeveloped World the standard of living is already so low, and the rate of increase so small (3 per cent annual economic growth rate, 2 per cent population increase, net standard of living increase of only 1 per cent) that we can only ask that they focus on the problem of population numbers.

The contribution we can make in terms of resource and energy consumption can, broadly speaking, be through the adoption of 'green' policies that are presently taking centre stage. We must support, emphasize and extend them.

But many of the current campaigns that are considered ecologically friendly are in fact somewhat trivial in nature and in some cases of doubtful validity. The skilful and energetic propaganda to save waste paper and collect glass bottles, for example, is a massive effort deployed for relatively small and not well-proven benefits.

Rather we must deploy our persuasive skills into fields where we can have a significant and telling impact. Advances are already being made in new high technology areas such as the design of the motor car (engines, tyres, roads), and in electricity power generation. But these advantages are very often outweighed by extravagances such as buying bigger cars and turning up the central heating control. Industry and commerce are not vigorously encouraged to save energy – though they could be through a taxation regime focused on energy-reducing practices. And society in general could concentrate its efforts less on recovery of spent resources (though a laudable cause in itself), more into persuading people to prevent the creation of waste in the first place.

All this, however, is a digression for this book as the main theme is intended to be population numbers.

One suggestion to reduce numbers is to adopt the policy of encouraging one child per family – but no politician would grasp this nettle. But what if, as a start, we mounted a campaign to encourage a maximum, not average, family size of two? The number of births in the West would fall by 3 million a year – a sizeable contribution to eliminating the problem. In consumption and pollution terms, this would be equivalent to a reduction of 100 million people in the Underdeveloped World and, incidentally, roughly matches the efforts to be asked of them.

However, the problem is that even if we managed to reduce the number of children to two on average, already we

are just below two in the Developed World, numbers would keep rising over the next decade or more because the number of young women coming to puberty as a percentage of the total population is presently increasing. The Underdeveloped World will experience this phenomenon much more than the Developed World, though the trend will be common to both. Sooner rather than later, the target of fewer than two children must be promoted.

Another possible step might be to persuade, not coerce, young women to delay having their first child until they are older. On average in the West, the first child is born when the mother is aged 25. The West's commitment to doing something in tandem with action by the less fortunate areas of the world would be visibly reinforced if we encouraged women to delay motherhood: for each year postponed a one-off reduction of 15 million births could be hoped for.

Chinese experience is interesting. During the 1970s,

population control efforts were focused on the age of marriage. In the 1960s it was 20 for males and 18 for females, but in the 1970s it was raised to 27 and 25. The fall in birth rate was dramatic: from 34 to 18 per 1,000. It has been estimated that half the fall was due to later marriage and half to social pressures for fewer children. The one child policy of the 1980s is having a much smaller impact and its serious shortcomings need to be dealt with.

In the European Union, around half a million unwanted children are born each year, most of them to women well below the age of 20. The whole of this area is fraught with emotion, but it does seem sensible to use our educational capability to teach contraceptive methods and human relationships at an early age. Good contraceptive education might help avoid such an unhappy outcome for young girls.

What else can be done? Should not every government, Western and Third World, have in its cabinet a minister whose sole duty is to be concerned with population matters? This is already the case in India. This portfolio would embrace collecting and publishing appropriate statistics, ensuring that the population is made aware of all of the implications of the collected data, setting objectives for the various facets of population growth, persuading all governments to allocate funds for dissemination of information and other teaching techniques, and ensuring adequate manufacture and distribution of birth control methods. There are many voluntary bodies who could also get involved in the wider national effort. Efforts should be made to create dedicated ministerial posts at European Union and United Nations levels.

Then there is the question of governmental and charity aid directed to countries in need – currently around $60 billion a year. At present the Developed World gives aid equivalent to about one third of 1 per cent of the donor countries' gross national product. The figure was 50 per cent higher 30 years ago. So while we might be donating more in total, in percentage terms the amount is falling.

Would it significantly affect our economic well-being to increase our present donation of $60 per person per year by 50 per cent to $90? One possibility would be to make it conditional on the extra being spent totally on contraceptive measures, educational as well as physical. Then, could not the country that shows the biggest improvement in population reduction be given the reward of even more aid and thereby encourage others to seek to get the same benefits? Interestingly this 50 per cent increase would raise enough money to bring reproductive health to the whole planet and to bring on the development, education and changes in women's status essential to make a smaller family the rational choice.

In this proposal we meet one of our ground rules in its harshest form. What about those people in the Third World who are presently depending on receiving aid just to survive day to day? Any aid directed to contraceptive measures clearly detracts from help that could be directed to alleviating today's suffering.

It is frequently said that families in Underdeveloped countries have children because they need security in old age – a pension in human form. Could we not consider the almost crazy idea that the West seeks the political and organizational means to guarantee pensions to old people in the Third World? We might make payment conditional on the recipient not having more than one child.

The arithmetic is interesting; assume there are 800 million people entitled to a pension of, say, $250 per annum – that's half their present average income. That would cost the West $200 billion which, before you switch off completely, is just over 1 per cent of our GNP. The administrative and political difficulties are almost certainly insurmountable, but surprisingly the economic cost is less frightening than you may first think. Playing this wild and fanciful card is to bring home the need to explore imaginative ways of doing something to help the Underdeveloped World to solve the population problem – because it is our problem too.

New and radical thinking is required if we are successfully to handle the problem of too large a world population. Some people – probably many people – feel that the Developed World already gives large amounts of aid to the Underdeveloped World, and that no more should be given. The approach to the problem, therefore, is to improve the way in which aid is deployed, not to reduce it. We must not fall into the trap of believing that we are over-generous to the Underdeveloped World; we are not. But it is *in our own interests* to find solutions to population growth.

If we accept political and religious differences as being formidable and unbridgeable, then there is no alternative but to despair. That desperation must not set in, because the stakes are too high. If, on the other hand, we can persuade nations to place this whole subject centre-stage, then we are in with a chance. Only we, all of us, are the ones that can bring about the necessary change. This can be achieved in many ways, of which this book is but one small contribution to the most important debate on earth. Could we not also ensure that our political representatives have to rely on their record on population issues to win their seat at the next election?

Mass communication, TV and radio in particular, is an important weapon in our armoury. We in the Developed World have the means and skills to put over the message. Others do it. Religious organizations beam their message of hope and salvation around the world and there is evidence that converts are made; why not similar efforts in the field of population concern?

15 Final Thoughts

It is hard, if not impossible, to avoid worries felt across the world. The increasing globalization of society, the expansion of information technology and the feeling of the fragility of man in his environment are just a few. Then there is work insecurity, collapsing state authority and a wide range of moral dilemmas. All this is deeply depressing.

So over-population is hardly the only problem. It is, however, relevant, specific and one that mankind can solve. To this extent it is unique and if it is not dealt with, many other global concerns could become even more massively insoluble and overwhelming than they are at present.

If you want to reduce the threat of global warming, pollution of the great oceans or the cutting down of our forests, you cannot ignore their origin. Yes, rising economic expectation and activity has to be addressed, but so too has the question of population numbers – and this issue above all must be placed centre-stage. Economic expansion to meet rising standard of living expectations will demand vast industrial transformations that will take decades to achieve, and if nothing is done in the meantime about population numbers, the possibility of a successful economic outcome will be undermined. The two strands run together.

To elaborate, if the growth in the world population could be stopped overnight and if the target economic growth of 3 per cent per annum was achieved the general living standard of each individual would roughly double over the next twenty-five years, more in the Under-developed World and less in the Developed one; a mind-bending increase in terms of pressure on the world's resources and capacity to absorb pollution. Unhappily, on a per person basis, the benefit will only be about one-third of this unless the present rate of growth of the world population is curtailed.

The environmental consequences of a 3 per cent world economic growth target are dire. The problem we face is how can mankind continue to raise his economic well-being, aiming roughly to double his per capita living standard over twenty-five years, without the environmental consequences of a doubling of the overall total world economic activity. Clearly this can be achieved only by a reduction in population numbers, which can be brought about only by a birth rate of less than two per female; that is less than a one for one replacement.

The balance between population and the efficient use of resources has become badly out of kilter. Both aspects are important. Politicians and those claiming green credentials must be persuaded to strive for more equality of treatment between the two. The subject of over-population is not yet centre-stage, yet the action required to solve it is possible without any new or original research.

The means of disseminating information are available now, and the physical means for birth control are manufactured throughout the world. Better use should be made of modern sophisticated communications media, including the information highway, to allow us to enter each home and to persuade everyone that the time for action is now.

Those who talk, write treatises and give lectures on

topics covered in this book are often described as 'gloom and doom merchants' because their messages are never welcome. But something can be done if the problems are tackled now. We must achieve the objective of making anxiety about over-population one of the central planks of government. From that will come solutions.

You, the reader, must take up the cudgels. You can make governments in the Developed World sit up and take notice. Only you, the voter, can persuade your politicians that the problem discussed in this short book is of great moment. Our leaders have the difficult task of determining what is politically possible, and you, the voting citizen, can bring pressure to bear.

I hope this book has persuaded you that it is not some distant generation that may suffer the effects of over-population; it is affecting you now, and it will have profound implications in the future for your children. Do not be like television audiences who, when responding to some unhappy world catastrophe, say that 'something must be done' without quite knowing what should be done. Ask yourself 'What will be done on Monday morning?'

You are entitled to say that you do not fully know the answers to over-population – but you are also the one to demand, loud and clear, that the topic be made a central political theme. The truth is that unless we address the question, we will never find the answer. It is a race against time.

16   Summary of Statistics

How Many of Us Are There?

- Presently 5.7 billion and increasing by over 275,000 each and every day. Only 200 years ago, at the time of Napoleon, we numbered 1 billion.
- 80 per cent of the increase is in the Underdeveloped World.
- One third of that number is Muslim, one fifth is Catholic and one sixth Hindu.
- One billion people live in the affluent Developed World, three quarters of a billion in the Developing World and four billion in the Underdeveloped World.
- The West has an average annual income of $17,000 compared with $500 in the Underdeveloped World – a ratio of 34:1.
- China's forceful stance on late marriage and one-child families is working ...

...and the population is expected to increase by 50 per cent in 25 years.

What About Fuel Reserves?

- Energy is the ultimate measure of activity; the staff of life.
- Total world energy consumption is the equivalent of 8,000 million tons of oil, of which only around one sixth is non-fossil fuel based (nuclear, hydro-electric).
- Oil and gas could run out within 40 years or so – within the lifetime of many of us.
- Coal reserves are in hundreds of years.
- The Developed World consumes 10 times the energy per person of the Underdeveloped World and 25 times the electrical energy.
- A calculation made solely for dramatic purposes shows that it would require 4,000 large power stations to be built overnight to bring the Underdeveloped World up to the West's standards. It takes 10 years to build one station.
- To cope with a 100 million population increase each year would require a further 100 power stations to be built every year.
- Energy availability and power generation capability simply will not keep up, even allowing for renewable sources.
- To change the industrial base to wind and wave machines, harvesting machinery, and so on, would take decades to achieve – by which time the population pressure will have swamped the likely benefits ...

... and the population is expected to increase by 50 per cent in 25 years.

Is Global Warming a Threat?

- Global warming is a phenomenon caused by an increase in greenhouse gases, mainly carbon dioxide, which produce a heat protection blanket around the earth.
- Most of the increase in carbon dioxide is the result of man's profligate use of fossil fuels.
- The United States (population 230 million) produces 27 per cent of these gases and China (population 1,300 million) 8 per cent.
- The global temperature is expected to rise by 1°C over the next 25 years; in geological terms this is a staggering amount.
- The sea level is rising at half an inch a year or around 8 inches over the next 25 years with predicted catastrophic consequences.
- The 50 million people who presently experience the ravages of massive flooding each year will double in the next two decades; and a further 400 million will be seriously affected.
- 500 million more people will suffer hunger over the same period as a result of global warming.
- Preventative measures require to be measured in GNP terms because the sums involved are so vast.

… **and the population is expected to increase by 50 per cent in 25 years.**

How Fast Are Our Forests Disappearing?

- In the beginning 80 per cent of land was covered by forests; now it is 30 per cent and falling.
- There has been a tenfold reduction in the forest area per person over the last 200 years.
- Half the growing forests are in temperate zones, which

roughly correspond to the Developed World. A replanting programme is bringing them to the level of sustainability.
- Not true of the Underdeveloped World; an area of growing forest the size of Great Britain is destroyed each year. Reflect on the loss of 'lung' capacity, biodiversity *et al.*
- Most of this cleared forest area is consumed – indeed caused – by the demands of agricultural needs, which arise from population growth.
- A lesser requirement is space for urbanization. The demand for cleared space for urbanization is, however, not insubstantial; think of the ever-expanding mega cities of Buenos Aires, Calcutta *et al.*

... and the population is expected to increase by 50 per cent in 25 years.

Is There Enough Food?

- On average there is just enough food produced to feed the present world population – but it is extremely unevenly spread.
- Some 5 per cent of the world's population – 250 million people – is severely malnourished and consumes only 1,600 calories per day. A further 1.5 billion are significantly below the minimum nutrition required for a healthy and active life.
- The 1 billion people of the Developed World have an average calorie intake some 40 per cent higher than that required for a healthy and active life.
- Because of increasing numbers and the encroachment of the deserts and the sea, the fertile land available per person for crop growth in the Underdeveloped World has decreased by 30 per cent in only 2 decades.

- The total world food production is at present roughly static, possibly even faltering downwards.
- Fertilizer consumption is 70 kilograms per person in the Developed World and only 15 kilograms in the Underdeveloped World. To match the Developed World, an overnight tripling of fertilizer production is required; this scale of increase is impossibly large.
- The likely increase in population will swamp any increase in food productivity. It is claimed that genetic engineering can increase food availability mainly by reducing losses due to bacterial and fungal attack and so on, but the true ultimate factors are energy, water, nutrients and time; all in very short supply.

... and the population is expected to increase by 50 per cent in 25 years.

Is There Enough Water?

- The water required to feed a person for a day, including that necessary for growing food, is 200 gallons. This makes no allowance for commercial and industrial needs.
- The average rainfall worldwide is 4,000 gallons per person per day – but it falls very unevenly and, for the most part, in unpopulated areas. Rainfall in the Nile Valley is only 25 per cent of that in Europe.
- Two billion people, one third of the world's population, in 80 countries live with chronic water shortage.
- The Aswan Dam meets the needs of 10 million people; only some 5 years' population growth in Egypt.
- Only the rich can pipe water to wherever it is required. Generally, distances are in the order of thousands of miles and the needs in millions of gallons. The capital

and energy costs of moving water, some $2000 per person, exceeds the average income of the Under-developed World by a factor of 4.

- Distillation of sea water is also impossibly expensive, except in some cases where oil and gas are cheap.
- Esoteric ideas such as towing icebergs from the poles are unrealistic in terms of cost and the quantities required ...

... **and the population is expected to increase by 50 per cent in 25 years.**

What About Pollution?

- Usable land is being consumed and degraded to meet the insatiable demand of a rising population for expansion of industry, electrical generation and housing.
- Desertification is increasing by an area the size of Great Britain every year.
- The origin of atmospheric pollution and specifically acid rain is mainly the outcome of the West's burning of fossil fuels; a case of self-mutilation.
- Timber harvests could be increased by 25 per cent if we eliminated car exhausts and sulphur emissions.
- The West produces 5–10 billion tons of solid industrial waste each year, a half billion tons of hazardous waste and a half billion tons of domestic waste each year.
- The seas have already reached their sustainable fish limit of 100 million tons per annum.
- Aquaculture accounts for a further 10 million tons and is extremely unlikely to be able to expand to meet future population needs.

... **and the population is expected to increase by 50 per cent in 25 years.**

The Impact of the Motor Vehicle

- There are 700 million vehicles in the world, of which three quarters are in the West.
- They consume one third of world oil production.
- A huge increase is predicted within 25 years, just under 100 per cent in the West and over 100 per cent in the Underdeveloped World.
- The average American spends 67 days per year in his car, travels 7,500 miles and only achieves 5 mph on average.
- Congestion costs are substantial at 5 per cent and 10 per cent for fuel and increased maintenance respectively.
- In the West, some 5 per cent of GNP and 4 per cent of jobs are vehicle-based.
- Containment steps being considered include increased fuel taxes and restricting the building of new roads ...

... and the population is expected to increase by 50 per cent in 25 years.

Unemployment

- Labour rates in countries such as China and Korea are 20 to 30 times lower than in the West.
- The unemployment level in Europe is presently 15 million; an extra 700,000 per year will be added just on population growth alone.
- Reference to population control results in immediate loss of income by about one third to 'green' charities and a loss of votes to the politician.

... and the population is expected to increase by 50 per cent in 25 years.

Animals and Other Matters

- Intensive farming produces hens that cannot walk, pigs and calves confined to restricted spaces, and featherless chickens.
- Sixty million buffalo were killed in ten years.
- Concern continues about the survival of the tiger, panda, whale, koala and elephant.
- The annihilation of elephants and rhinos for their tusks and horns continues.
- The loss of biodiversity is shocking: 40 per cent of mammal species, 10 per cent of birds and 50 per cent of fish are threatened.
- The loss of amenities in unspoilt and remote regions of the world is increasing because of the growing interest in leisure pursuits such as walking, climbing and sailing.
- The exponential growth in complexity of our society is causing our legal and administrative systems to grind to a halt.

... and the population is expected to increase by 50 per cent in 25 years.